DON'T THROW IT ALL
AWAY

FRIENDS OF THE EART
WASTE REDUCTION AND

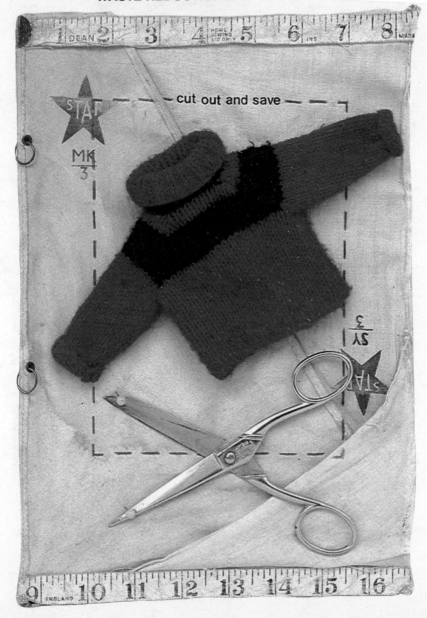

- - cut out and save - -

DON'T THROW IT ALL AWAY!
Friends of the Earth's Guide to Waste Reduction and Recycling
Author: Friends of the Earth
Editor: Sarah Finch
Illustrations: Pete Bateman
Design: David Caines

Thanks are also due to Michael Hirsh, Sara Huey, Penni Mawson, Liz Peltz, Neil Verlander, Peni Walker, Adeela Warley and Joanna Watson. Thanks to Carl Doyle at the Worx.

This publication has been partially funded by the London Boroughs Grants Committee.

Friends of the Earth
Friends of the Earth is one of the UK's leading environmental groups. It works locally, nationally and internationally and provides authoritative information on a wide range of international issues. It has an active network of over 300 Local Groups and a youth section, *Earth Action*.

Friends of the Earth is backed by more than 250,000 supporters in the UK. For information on how to join, or your nearest Local or *Earth Action* Group, or for a free copy of our publications or merchandise catalogue, please write to: Friends of the Earth, 26-28 Underwood Street, London N1 7JQ, enclosing a stamped addressed envelope.

Recycled Art
Pete Bateman was specially commissioned by Friends of the Earth to create twelve original artworks that reflect different aspects of recycling and waste reduction. Every picture has been constructed using discarded or waste materials to emphasise the message of this book. Pete is represented by The Indian Ink Company *Tel*: *(081) 960 8377*

FUNDED BY

November 1992
© Friends of the Earth
Published by Friends of the Earth Trust Limited
ISBN 1 85750 200 0

Friends of the Earth
26-28 Underwood Street
London N1 7JQ
(071) 490 1555

Friends of the Earth Scotland
Bonnington Mill
70-72 Newhaven Road
Edinburgh EH6 5QG

DON'T THROW IT ALL AWAY!

Contents

1. RUBBISH: A PROBLEM THAT WON'T GO AWAY

We throw away 20 million tonnes of rubbish from our homes every year in the UK - that's one tonne for each household.

For most of us, once our rubbish has left our homes it is 'out of sight and out of mind'. But when rubbish is thrown away it doesn't really go 'away'. It just goes somewhere else.

Between 80 and 90 per cent of our household rubbish ends up in landfill tips, where it rots down, generating methane gas and noxious liquids. Most of the rest is burned in municipal incinerators, creating polluting smoke, and ash which has to be landfilled.

We should also remember that the Earth's resources are not unlimited - everything we use has to come from somewhere. Wildlife habitats are damaged and pollution caused when resources are mined, harvested and refined to make everyday products like newspapers and cans.

More and more people now realise that everything we do has an impact on the environment. What we buy and choose not to buy and what we do with our waste all have an effect. Through our everyday choices, we can be a powerful force for change.

Friends of the Earth believes that it's time we consigned the 'throwaway society' to the dustbin of history. We hope that this booklet will show you some of the ways in which you can help to play a part in taking the pressure off the planet.

2. FRIENDS OF THE EARTH

Friends of the Earth is committed to building public understanding of environmental issues and promoting protection of the environment.

Since the early 70s, we have encouraged people to do what they can to 'help the Earth fight back'.

Waste and recycling issues have concerned us from the earliest days. In fact Friends of the Earth's first ever public 'action', in May 1971, was to dump 1,500 bottles on the doorstep of Schweppes' headquarters, to protest against the company's decision to phase out returnable bottles. We have continued to highlight the wastefulness of the 'throwaway society' and to promote waste reduction, reuse and recycling ever since.

In 1989 Friends of the Earth launched the three-year Recycling City project. Four areas - Cardiff, Devon, Dundee and Sheffield - took part, testing different methods of collecting and sorting recyclable material. Recycling City helped make people aware of the enormous potential for recycling. It also identified some of the barriers preventing local authorities achieving higher levels of recycling.

The situation has changed enormously since Friends of the Earth first started 'talking rubbish'. Now most people are keen to recycle their rubbish and the Government has set a target of recycling a quarter of household waste by the end of the decade, a big increase on the tiny proportion that is recycled at present.

In the years to come, Friends of the Earth will continue to research and publicize the problems caused by our throwaway culture, and to promote the message "Don't Throw It All Away".

3. WASTE: THE FACTS

■ What do we Throw Away?

As well as the 20 million tonnes of household rubbish we throw away from our homes, another 460 million tonnes of solid waste are disposed of annually from shops, offices, factories, farms, building sites, mines and quarries.

■ Household Rubbish

Household waste only makes up 4 per cent of the solid waste produced in the UK each year. But it has attracted more public interest than any other type of waste, because everyone produces household rubbish and so everyone has some responsibility for the problems caused by its disposal.

The Government has analysed the waste people produce. They found that the average dustbin contains:

paper and cardboard 33%
glass 10%
plastic 7%
ferrous metal 7%
textiles 4%
aluminium cans and foil 1%
compostable material, ash and dust 38%

A third of the rubbish in the average dustbin is packaging.

As well as the rubbish we put in our dustbins, 4 million tonnes of household waste are disposed of at 'civic amenity sites' (town tips). This waste includes bulky furniture and garden waste.

■ What Happens to our Rubbish?

Between 80 and 90 per cent of household and industrial waste is buried in landfill tips. About ten per cent of domestic and commercial waste is burned in municipal incinerators.

Only a very small proportion of the waste we produce gets recycled. Friends of the Earth's research has found that only between two and a half and four and a half per cent of household rubbish gets recycled.

This compares to around 10 per cent in Germany, 16 per cent in Australia, 20 per cent in Finland and up to 40 per cent in Japan.

4. THE ENVIRONMENTAL IMPACTS OF WASTE

Our 'throwaway' culture means that we use up large quantities of raw materials and energy. Burying and burning all our waste creates serious pollution problems.

■ Wasted Resources

Throwing so much away is a senseless misuse of the Earth's precious resources! One way to save resources is to use waste itself as a raw material. Even better is to waste less in the first place.

Some raw materials are 'finite' - which means that reserves will simply run out if we continue to squander them to make throwaway goods. Recycling products made from finite resources like plastics (made from oil) and aluminium (made from bauxite) can help to conserve supplies. Reducing unnecessary packaging and making things to last also help.

Reducing and recycling waste also relieves the environmental damage done to wildlife habitats when raw materials are mined, quarried or 'harvested'.

> Every tonne of recycled glass saves 1.2 tonnes of sand, soda ash and limestone.
>
> Recycling one tonne of steel saves 900 kilogrammes of raw materials.

It takes 10 to 18 trees to make one tonne of paper - so using less paper and recycling waste reduces the need for intensive forestry plantations and for felling trees in ancient natural forests.

■ Wasted Energy

Using scrap materials to make new products takes less energy too. Saving energy helps to reduce all the problems associated with energy generation: acid rain, smog, radioactive pollution from nuclear accidents and waste, the flooding of valleys for huge hydro-electric power schemes and the threat of climate change.

Potential energy savings from using scrap materials instead of virgin ones:

Glass	22%
Paper	70%
Aluminium	96%
Steel	74%
Polyethylene	97%

■ Pollution

Disposing of all the waste we create is a problem in itself. Both burying and burning waste can cause hazards to human health and to the environment. It is not only 'toxic waste' that is a hazard: household waste can also produce polluting liquids and explosive gases when buried in the ground and harmful fumes when incinerated.

Thousands of landfill sites around the country are generating polluting gas and liquids and threatening to contaminate drinking water supplies.

We are running out of sites where waste can be landfilled, especially close to cities where most of the waste is produced. So waste has to be transported further and further away from its source, adding to the environmental damage caused by its disposal.

5. REDUCE, REUSE, RECYCLE

The best way of reducing the problems caused by waste is to produce less waste. Where waste cannot be avoided, it should be reused or recycled wherever possible.

■ Waste Reduction: Throw Away the Throwaway Society

We are producing more and more waste. Between 1985 and 1989 the amount of household waste produced per person in the UK rose by 5 per cent, whereas in some other countries, including Canada and Norway, the amount of waste each person produced actually fell.

Friends of the Earth has asked the Government to set targets for **reducing** waste and to develop strategies for meeting them.

We could all create less rubbish by refusing to buy goods with unnecessary packaging and by buying clothes and household goods that are designed to last several years rather than a few months. Industry too can adapt its manufacturing processes so that they are less wasteful.

■ Reuse: Bring Back the Bring Back

Reusing something takes less energy and creates less pollution than reprocessing it to make a new product.

Many items - like clothes, furniture and household goods - get thrown away simply because they are old or broken. They could be repaired, refurbished and reused.

Products like food, drinks, household cleaning products and toiletries, could be sold in reusable containers instead of one-trip packaging. Studies have shown that reusable packaging systems use fewer materials and less energy than one-trip packs and create less pollution and waste. Friends of the

Earth is calling on the big supermarket chains to introduce reusable packaging schemes.

■ Recycling: Once is Not Enough

If waste cannot be reused it should be recycled. Over half of the rubbish in our dustbins could be recycled, but only between two and a half and four and a half per cent of it is.

Glass, plastic and metals can be melted and reshaped. Paper can be pulped and made into new paper. Organic waste can be composted. Textiles can be unwoven and respun into new cloth. But the enormous potential is not being met.

In 1990 the Government set a national target - to recycle 25 per cent of household waste by 2000. This is a massive increase on the amount we manage at the moment!

Local authorities will play an important part in making sure we reach this target. The Environmental Protection Act 1990 requires all local authorities which collect waste to write a Recycling Plan, setting out how they intend to recycle waste from homes and shops in their area.

These plans should have been submitted to the Government by August 1st, 1992. However fewer than half of the local authorities who should have prepared plans actually submitted them by the deadline. And only just over a third told Friends of the Earth that they were planning to meet or exceed the 25 per cent target.

You have a right to see your local council's Recycling Plan. Local authorities will be reviewing their plans in the coming years, and they should take residents' views into account. So let the council know if you think it could do more to increase the amount of waste that is recycled locally.

■ Recycling Collections

Councils around the country are using different methods to collect and sort material for recycling.

Most councils collect material by providing banks or recycling centres where people can bring their recyclables. These are known as 'bring' or 'bank' schemes because people have to bring material to the banks.

Some councils collect recyclable material from people's homes. Recyclables are collected separately from the rest of the refuse - in different dustcarts or in special dustcarts with separate compartments for each material. These are known as 'door-to-door', 'kerbside' or 'collect' schemes, because the material is collected from each household.

Door-to-door schemes are more expensive to set up in the short term. However, comparisons show that a higher proportion of the available material is sent for recycling using door-to-door schemes than where bring schemes are used. In the longer term, door-to-door collections may also lead to financial savings.

The London Borough of Sutton has 33 mini recycling centres - that's one for every 4,983 people.

Glass and paper are collected at most of the sites, cans at 5 sites, rags at 9 sites, and batteries at 12.

In 1991/92 more than 10 per cent of the total household waste produced in the Borough was collected for recycling - just over 6 per cent through the mini recycling centres and the rest through a door-to-door collection.

The Royal Borough of Kensington and Chelsea has tested door-to-door collections and found that they are popular with residents - 74 per cent of the people living in the trial areas took part and 15 per cent of the waste in the area was collected.

So the scheme is being expanded, and from April 1993 all the householders and businesses in the Borough will be able to put plastic, glass, paper, cans and rags out for recycling alongside their ordinary refuse.

Residents will be asked to sort their rubbish and put recyclable materials into a carrier bag and nonrecyclable waste in the usual dustbin or bin liner.

Specially designed 'split back' lorries will do the rounds, with recyclables put in one side and rubbish in the other.

The recyclable materials will then be sorted and sold on to companies for recycling.

The Council estimates that the door-to-door collection will enable it to meet the 25 per cent recycling target, and to save £250,000.

Leeds City Council has been running door-to-door collections since 1990.

The collection currently covers 12,000 households; it will be expanded to cover 28,000 households in early 1993.

Householders are provided with two wheeled bins: a green one for packaging materials, textiles and newspapers, and a brown one split into two sections for organic materials and non-recyclables (there are larger bins for blocks of flats).

The green bins are emptied one week and the brown ones the next. Glass is collected through a network of bottle banks (1 set per 400 households).

A survey conducted in 1991 found that 45 per cent of the rubbish collected from households was being diverted for recycling on a regular basis.

Britain - A Nation of Frustrated Recyclers

A January 1992 Mintel survey found that although 94 per cent of people think recycling is "very" or "fairly" important, about a third (32 per cent) had never taken anything for recycling and two thirds (64 per cent) usually threw all their rubbish away.

The main reason given was that there were not enough recycling facilities. 37 per cent of the people surveyed did not have a recycling collection near them and 49 per cent thought that more facilities were needed.

■ Overcoming the Barriers to Increased Recycling

Money

Many local authorities expect to have difficulty meeting the Government's recycling target in their area because they do not have the money to invest in recycling collections.

One major reason why recycling seems expensive for local authorities is that the traditional option - landfill - is very cheap. Some local authorities say it only costs them £2 or £3 a tonne to landfill waste. It is cheap because the environmental standards that landfill operators have to meet are low and badly enforced. In contrast, in the USA landfilling household waste costs over $100 a tonne.

About one third of the rubbish in the average dustbin is packaging. Many local authorities would like to see the packaging industry contributing towards the costs of collecting this waste for recycling.

The Government's top recycling expert has estimated that it will cost local authorities about £200 million a year to set up and run the type of recycling collections that are needed to meet the 25 per cent recycling target. The Government will need to ensure that local authorities have access to the funds they need if it is serious about meeting its target. It could also enforce tougher standards for landfilling waste, to make the price reflect its true environmental costs and therefore make comparisons with the cost of recycling fairer.

Markets

Some local authorities are not planning to collect recyclable materials because they cannot find a company which will recycle them.

Finding a market for recyclable materials is important to provide an income for local authorities and to make sure that the collected matterials actually get recycled.

There are a number of measures that the Government could adopt to develop the markets for recyclable materials.

These include:-

a law specifying the minimum percentage of recycled paper to be used in newspapers;

a law requiring Government bodies to use recycled materials;

price incentives to encourage people to buy recycled products.

■ Waste Disposal - Dumping the Dross

Even if we take all possible measures for waste reduction, reuse and recycling, there will always be some waste and its disposal will always have an impact on the environment.

However, we can substantially reduce the problems by ensuring that waste disposal is controlled by strict laws to keep environmental damage to an absolute minimum, and that the regulations are properly enforced.

6. KITCHEN AND GARDEN WASTE

Around a third of household waste consists of compostable organic material like fruit and vegetable peelings, stale food, garden waste, finger nail trimmings and bedding from the pet's cage!

Compostable material also makes up between a fifth and two thirds of the waste taken to civic amenity sites, depending on the time of year.

Altogether, an estimated five to seven million tonnes of organic waste are thrown away by householders each year. This could produce between four and six million tonnes of compost.

When organic waste breaks down in **anaerobic** conditions (when oxygen is absent) it produces methane gas, which is highly explosive and which contributes to the greenhouse effect (see explanation right). If organic waste breaks down in **aerobic** conditions (when oxygen is present) it produces carbon dioxide, also a greenhouse gas but a less potent one than methane.

What is the Greenhouse Effect?
The greenhouse effect is a natural phenomenon, whereby heat from the sun is trapped by a blanket of 'greenhouse gases' in the Earth's atmosphere.
Without the greenhouse effect life as we know it could not exist, as the Earth would be too cold.
However, human activities are now increasing the concentrations of a number of gases, including carbon dioxide (largely from the burning of fossil fuels - like coal and oil - in power stations and motor vehicles) and methane.
This is expected to increase the greenhouse effect, which may lead to changes in the Earth's climate and cause sea levels to rise.
Unless we take steps to cut our emissions of greenhouses gases, we may experience serious problems by the year 2030, including frequent floods in some areas, drought, famine and starvation and millions of environmental refugees.

There isn't much oxygen inside a landfill tip, so buried organic waste generates methane as it rots, posing an explosive threat to nearby buildings and contributing to the threat of global climate change. It also contributes to the production of 'leachate', toxic liquid which can leak out of the tip and pollute water.

Kitchen and garden wastes can be recycled by composting, to produce a good soil conditioner.

If you have a garden you can compost your waste at home. Put a good mix of compostable waste in a container (home-made chicken wire or wooden bins are just as good as most manufactured ones), and make sure the material stays moist (but not wet) and has enough air. Turn the compost over once a month if possible. Remember that the process is carried out by living creatures - provide them with water, air and a good mixture of raw materials and you can't go wrong!

A survey in Wye, Kent estimated that 15 to 20 per cent of households in the village compost their organic wastes on a compost heap at home.

If you haven't got a garden you can still compost your kitchen waste using a 'worm bin', and use the compost for pot plants or give it to friends with gardens. You need a plastic dustbin and some special compost worms (also known as Brandling worms, manure worms or *Eisinia Foetida*, these are obtainable from established compost heaps, anglers shops or in inexpensive kits). Feed the worms on your food scraps and they will process them into compost for you.

Because organic waste makes up such a large proportion of household waste, many local authorities are setting up schemes for recycling it, to reduce their disposal costs and to help towards meeting the 25 per cent recycling target.

Some local authorities are encouraging residents to compost their organic waste at home.

The London Borough of Sutton is offering a choice of different types of composting equipment, including worm bins, to householders at a subsidized price. Just over 7,000 households have taken up the offer.

Others collect compostable waste from households door-to-door and compost it centrally.

The City of Dundee District Council collects kitchen and garden plant waste from around 5,000 households, who put it in specially designed wheeled bins. The waste is composted with garden waste which is brought to recycling centres, plant wastes from parks, and a range of types of compostable waste from commercial premises.

Some local authorities compost organic waste from civic amenity sites and other sources such as fruit and vegetable markets, parks and farms.

The North London Waste Authority (NLWA) composts garden waste that is brought to civic amenity sites together with organic waste from parks and commercial sources. The compost is sold to the public and local authorities for use in parks and gardens, and to garden centres and commercial growers.

Some councils treat organic waste by **anaerobic digestion**. This is not composting. Anaerobic digestion produces a 'digestate', which can itself be composted, and methane-rich bio-gas, which can be burned to generate energy.

For Peat's Sake!

Composting can help to save our peat bogs. Peatbogs are a unique and important habitat for rare plants and animals, like the beautiful Bog-Rosemary, the insect-eating sundew and the rare damselfly - but they are being destroyed to produce peat for gardeners and the horticulture industry.

Every year in the UK peat companies extract one million tonnes of peat, mostly for use in our gardens. 96 per cent of British raised peatbogs (the most ecologically important type) have been destroyed, and the remaining 4 per cent are under threat from commercial peatcutting for horticulture.

Gardeners can help to prevent the destruction by refusing to buy peat. Alternatives to peat are widely available - and one good alternative is compost made by gardeners from household waste.

7. PAPER

Paper plays a vital role in our society, but its production and disposal can cause a great deal of unnecessary environmental damage.

In some parts of the world, especially in Canada and the United States, ancient natural forests are being clear-felled and turned into paper. In other areas, artificial forests are being planted to produce pulp for paper. Ancient forests in Scandinavia, peatbogs in Scotland and Finland, and tropical rainforest in Indonesia are being destroyed to make way for plantations to supply wood to the paper industry.

Artificial forests are a poor replacement for natural ones, and can't support the same variety of plants and animals that lived in the area before. Forestry plantations can also disrupt the flow of water in the region and cause the water table to drop and wells to dry up. The fertilisers and pesticides used can pollute streams and contaminate drinking water supplies.

Pulp and paper mills discharge polluting effluent into rivers and sewers. This pollution can harm wildlife and have devastating effects on the environment.

Finland and Sweden show a good example of the damage intensive forestry can do. Between them these two countries supply a third of the pulp and paper we use in the UK.

In both countries natural forests have been replaced by forestry plantations. In Finland's forests 138 wildlife species, including the European beaver, have become extinct since 1985. 217 more are in *"immediate and serious"* danger of extinction. The threatened species include the Arctic fox, the lynx, the northern brown bear, the ringed seal and the oak mouse. In Sweden's forests, an estimated 140 species of plants and animals are threatened with extinction.

Paper buried in landfill sites contributes to the production of methane, the explosive greenhouse gas.

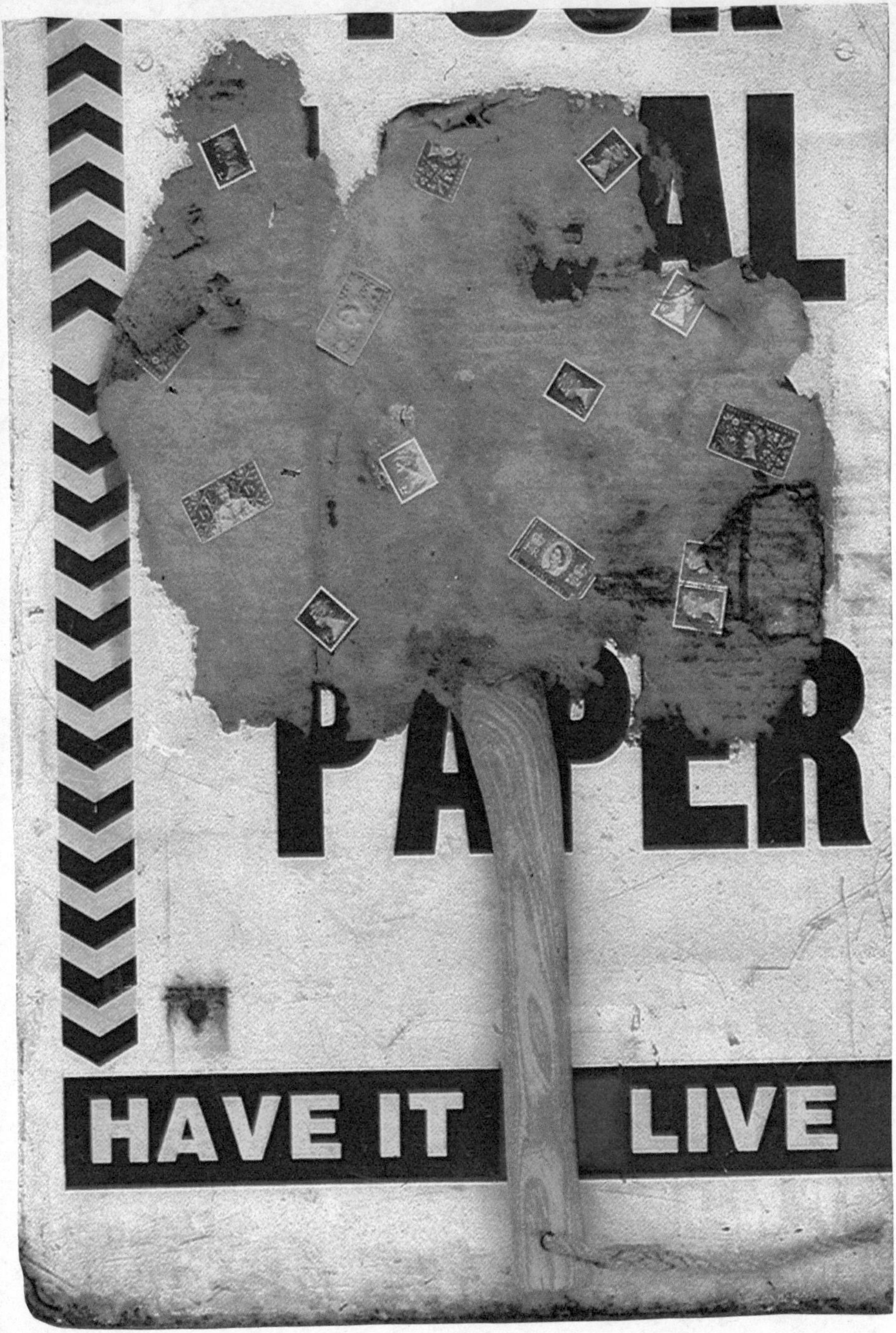
HAVE IT
LIVE

In order to alleviate all these problems we need to use less paper and to recycle more of what we do use.

In 1991 a third of the paper used in the UK was being recycled. It is thought that the percentage of paper from households that gets recycled is much smaller than this.

■ What kind of paper should I buy?

About two thirds of the paper we throw away at home is low grade (like newspapers, magazines and junk mail). But most of the recycled paper products we buy are made from recycled high grade waste (like office stationery and waste from paper mills and printers). If we are to increase paper recycling we need to use a lot more low grade waste paper.

Everybody can help to ensure that more low grade waste gets recycled by buying stationery and toilet paper made from old newspapers instead of high grade waste or virgin wood pulp.

Old newspapers can be recycled into new newspapers. Already some local newspapers are printed on 100 per cent recycled paper. But in 1990 just 26.8 per cent of the paper used by British newspapers was made from recycled fibre.

Conning the Green Consumer?
Sometimes you see paper described as *"environmentally friendly"*. This usually means that it has not been bleached with chlorine gas bleaches, the most harmful of the bleaches used in paper making. It does not necessarily tell us anything about the company's forestry practices, or about any other aspects of the manufacturing process. Sometimes paper companies boast that they *"plant more trees than they cut down"*. This may be true, but remember that there's a big difference between natural forests and artificial forestry plantations - from the point of view of the wildlife that lives there!
If you want to help protect the environment, buy recycled paper and don't be misled by 'green' labels.

8. GLASS

Glass is the material which is most commonly collected by councils for recycling in the UK. Every local authority now has at least one bottle bank.

Recycling glass saves raw materials and energy, but even greater savings are made if glass containers are reused.

Glass jars and bottles can be cleaned and refilled many times. No-one thinks it strange that milk bottles are reused - they are filled at dairies, delivered to households, then returned empty to the dairies for refilling up to a hundred times. But sadly reusable containers are less often used for other kinds of product.

Some shops and off-licences sell drinks in returnable bottles but none of the big national supermarket chains take empties back for refilling. The rise of the supermarkets has led to a decline in the use of returnable bottles.

The supermarkets claim that people would not bother to bring bottles back. However in May 1991 Friends of the Earth surveyed over 35,000 shoppers all over the country and found that 84 per cent of people would be willing to take bottles back for refilling if supermarkets set up schemes to collect them. Friends of the Earth presented the results to the supermarkets and asked them to 'bring back the bring back'.

In 1977, 60 per cent of fizzy drinks and beer in the UK was sold in returnable bottles. By 1987 this had dropped to 19 per cent and 23.3 per cent respectively. In Denmark, all fizzy drinks and 99.4 per cent of beer are sold in returnable bottles.

Glass bottles and jars which can't be returned should be recycled. Most bottle banks collect clear, green and amber glass separately. Most of the glass manufactured in the UK is

clear, but we import a lot of wine and beer in coloured glass bottles from other countries.

At present there are around 7,500 bottle banks in the UK - one for about every 7,500 people. In 1992 around 26 per cent of glass (including glass containers and flat glass) got recycled.

This figure could be higher if there were more bottle banks, or if glass was collected door-to-door. In Switzerland and the Netherlands a single bottle bank serves fewer than 1,500 people on average and over 50 per cent of glass is reclaimed. In April 1990, the door-to-door collection scheme in Sheffield collected 62 per cent of all the available glass waste.

9. CANS AND FOIL

Metals make up 8 per cent of household waste. Most of this is packaging - in the form of cans and foil.

Two types of cans are used in the UK: steel and aluminium. Food cans and half of all drinks cans are made of steel, often coated with a layer of tin. The remaining drinks cans are aluminium. (Most steel drinks cans have aluminium tops, which allow ring-pulls to be used.) It is easy to tell steel and aluminium cans apart because steel is magnetic and aluminium is not.

Foil is made from aluminium - foil represents nearly 40 per cent of all aluminium packaging in the UK.

The extraction and smelting of bauxite (the raw material from which aluminium is made), iron and tin are polluting processes. These valuable materials, and energy, can be saved if used cans and foil are recycled.

Everyone can help to reduce environmental damage by buying food and drinks in refillable containers, rather than non-refillable cans, where possible. And make sure that those cans that you do buy get recycled.

Bauxite, the ore from which aluminium is made, is often mined in tropical rainforest areas including Brazil, Guyana, Guinea, Indonesia and northern Queensland in Australia. The forest is cleared and a highly polluting red mud is left behind after refining.

In 1991, 11 per cent of aluminium cans and 10 per cent of steel cans were recycled. Some other countries achieve much better recycling rates, for instance in Germany and Switzerland 50 per cent of steel cans are recycled and Sweden recycles 90 per cent of its aluminium cans!

In the UK 26 local authorities - together covering 8 million people - extract steel cans and other ferrous metals from mixed waste by passing the rubbish under a giant magnet.

In July 1991, fewer than half of all local authorities had can banks. Because merchants pay a higher price for aluminium than for steel, there are more collections for aluminium cans than for steel ones. Save-a-Can banks take both kinds of can; when the bank is full the cans are sorted and recycled separately, but in August 1992 there were only 784 Save-a-Can banks in the UK to serve all those people whose councils don't have magnetic extractors - that's about one for every 66,000 people!

Can Ban!
In Denmark it is illegal to sell drinks in cans. The Danish Government passed this law in 1985 to help make sure that refillable bottles are used as much as possible. British can makers helped to take Denmark to the European Court of Justice, claiming that the can ban was a barrier to trade within the Common Market. But in 1988 the Court ruled that the Danish law was acceptable because protecting the environment is more important than free trade.

Each household throws away one and three quarter kilos of foil each year. Foil is made from different alloys from those used to make aluminium cans, and should not be put into can banks. Some charities collect foil for recycling. The foil manufacturers are currently running pilot foil collection schemes in five areas. They hope to expand to fifty areas in 1993 and 100 in 1994.

In December 1991 a large aluminium can recycling plant opened in Warrington. It can recycle 50,000 tonnes of cans a year. But most of the cans it recycles are imported from other countries - because we don't collect enough cans in Britain!

10. PLASTICS

7 per cent by weight and 20 per cent by volume of typical household waste is plastic.

Plastics are made almost entirely from oil. Recycling plastic helps to conserve the Earth's oil reserves; it also leads to considerable energy savings.

The production and disposal of plastics can cause environmental problems. Plastic factories discharge polluting wastes into rivers and sewers. PVC, a type of plastic that is widely used in packaging, is a major source of chlorine in waste. If chlorine is incinerated it contributes to the production of dioxins (highly toxic chemicals) and acid rain. These problems can all be alleviated by using less plastic and reusing and recycling more.

35 per cent of the plastics we use is for packaging. Plastics do not smash like glass or ceramics, so plastic containers could be reused time and time again. Dairies in the US and Canada have begun using refillable plastic milk bottles. Dairies in Sweden also use reusable plastic milk bottles, up to 100 times each. In this country, the Body Shop sells toiletries in refillable plastic bottles.

Plastic containers that can't be reused should be recycled. Unfortunately the plastics industry has put little investment into setting up schemes for reusing and recycling plastics. In July 1991 fewer than one in ten local authorities in England and Wales provided plastic collection points.

Overall, only about 4 per cent of used plastic is recycled in the UK. But most of this comes from commercial sources and less than one per cent of household plastics get recycled.

The potential is much greater than this. The Government's researchers have estimated that up to 70 per cent of the plastic in household waste could be recovered, through plastic banks and more door-to-door collections. The door-to-door collection in Sheffield recovered 68 per cent of the available plastics in April 1990.

One of the reasons why plastic recycling is so undeveloped is that a large number of different polymers (types of plastic) are in use. Some products are made from several different polymers. Mixed plastic can be recycled to make certain types of product, such as fence posts and traffic cones, but the polymers need to be sorted and recycled separately in order to make high quality products.

It would be much easier to recycle plastics if the plastics industry set up more sorting facilities for collected plastics, if each product were made of one polymer only, and if all products were labelled so that people could identify and sort them.

'Degradable Plastics' - A Red Herring
Some products are made of so-called 'degradable' plastic. With very few exceptions, these are not truly biodegradable. They do not break down into natural substances, they just fragment into smaller pieces of plastic. Degradable plastics do not solve the environmental problems associated with the use of plastics, and they may make some of them worse. Most degradable plastics are still made from oil, so they do not help to conserve the Earth's limited oil reserves. If a plastic is labelled 'degradable' people are misled into thinking that the best option is to throw it away, instead of reusing or recycling it. Degradable plastics can also make plastics recycling more difficult. There is no evidence that it is any less polluting or energy-intensive to manufacture 'degradable' plastics than ordinary ones. And toxic substances in the plastic, such as heavy metal dyes, may be released into the environment in an uncontrolled way when the plastic breaks down. Instead of using 'degradable' plastics to make throwaway bags and bottles, the packaging industry could help the environment more by cutting out unnecessary packaging and by making more containers reusable.

11. TEXTILES

Every year we throw away three quarters of a million tonnes of textiles, ranging from valuable second-hand clothes to carpet offcuts and dirty rags. It is estimated that less than a quarter of these discarded textiles are reused or recycled.

A large quantity of second-hand clothes already get reused, through charity shops and jumble sales. And many of the clothes that are collected by charity shops, rag merchants and local authorities are exported to developing countries to be reused.

Textiles that can't be reused can be recycled. Some fibres - wool, cotton and acrylic - can be respun to make new textiles for coats and blankets. Other textiles are recycled into products like roofing felt, wiping cloths and stuffing for upholstery.

Some local authorities have 'rag banks' to collect textiles. But in July 1991 less than a fifth of local authorities provided collection points for textiles.

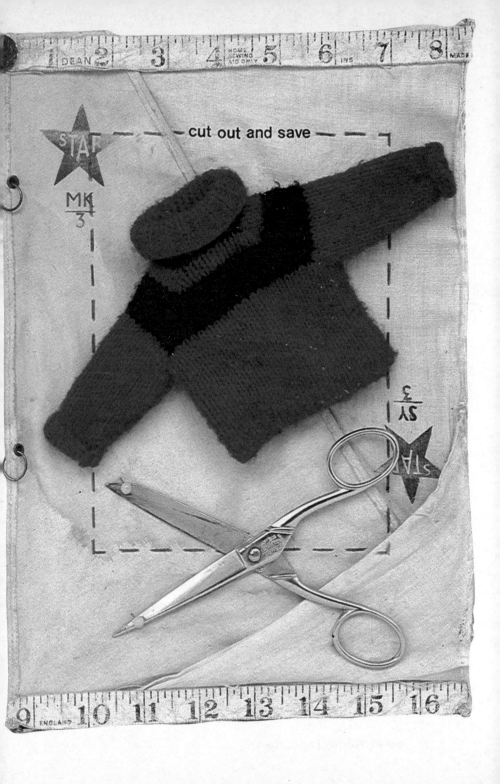

12. FURNITURE AND HOUSEHOLD GOODS

Many household articles which could be reused are thrown away every year, simply because they are old or out of fashion or because they need to be repaired.

Where possible electrical appliances, furniture and household goods should be repaired and reused instead of thrown away. Some local authorities run special collections for items of this sort and there are many other groups which renovate used furniture for reuse.

About a quarter of a million tonnes of steel scrap are thrown away annually in the form of 'white goods' (fridges, freezers and washing machines). White goods which cannot be repaired can have their metal content recycled. The steel industry buys scrap of this kind from local authorities, dealers and scrap metal merchants.

Old fridges and freezers contain chlorofluorocarbons (CFCs), both as a coolant and in foam insulation. If the CFCs are allowed to escape when the appliance is scrapped, they will damage the Earth's protective ozone layer. It is relatively easy to recover CFCs from coolant but in the UK there is no legal requirement for councils to carry out CFC recycling. In some countries, including the US, Sweden, Canada, Australia and the Netherlands, CFCs must be recycled by law. The technology to recover CFCs from foam insulation does exist but at present it is not in operation in the UK.

Friends of the Earth's research has revealed that barely 3 per cent of the CFCs in the fridges disposed of each year from people's homes is actually recovered by local authorities.

If your family is disposing of an old fridge or freezer, make sure that the CFC coolant will be removed and disposed of safely. If you are buying a new fridge or freezer the manufacturer or retailer may collect your old one and guarantee to remove the CFC coolant safely. If this service is not on offer inform the company that you will buy your appliance elsewhere.

13. BATTERIES

Batteries only make up a small amount of household waste - about 0.1 per cent. However they can cause serious pollution problems. Most of the batteries sold in the UK contain heavy metals, which are highly poisonous and very persistent in the environment.

Many local authorities collect car batteries at civic amenity sites. But there are no facilities in the UK for recycling smaller household batteries, and nearly all of them end up being thrown away.

Batteries can contain a variety of metals, including cadmium, mercury, lead, nickel, silver oxide, manganese and lithium. All of these are potentially harmful to the environment and/ or human health.

Some battery manufacturers accept their own rechargeable batteries back at the end of their lives; these are sent abroad for recycling.

Use elbow grease or mains electricity rather than batteries where possible - it takes many times more energy to make a battery than you will ever get out of it.

A European law has been passed which will ban batteries with heavy metal concentrations over certain levels by January 1994. It will require separate collection and disposal of some types of batteries and better labelling and information for consumers.

14. OIL

Motor oil is used to lubricate engines. While in use it collects lead and other poisonous additives from petrol.

If oil is poured down a drain which discharges directly into a river it can cause serious environmental damage. If it is disposed of via the sewerage system, it can stop sewage works working properly.

It is illegal to pour used motor oil down the drain or to dump it in soil. However it is estimated that three quarters of all do-it-yourself mechanics do just that. An estimated 100,000 tonnes a year of waste oil is unaccounted for in the UK, mostly from do-it-yourself oil changes.

Large quantities of used oil are burned in space heaters in garages. Burning oil in this way can release toxic heavy metals into the atmosphere. By law, anyone who uses a space heater must obtain a licence from the local authority, but it is believed that many space heaters are used illegally.

Instead of being dumped or burned, used oil should be reused. Many of the impurities can be removed from used oil, which can then be used as engine oil or fuel.

In July 1991 fewer than half of all the local authorities in England and Wales provided any facilities for collecting used oil. Many garages have collection tanks for oil.

15. WOOD

Wood makes up between 4 and 10 per cent of the waste taken to civic amenity sites, depending on the time of year.

Like other kinds of organic waste, wood buried in landfill sites rots down, generating the explosive greenhouse gas, methane.

Timber from demolished buildings can be reused in new buildings, or to make fences, pallets and cases. Reusing timber in this way could help reduce the demand for wood from unsustainable sources including tropical rainforests.

Small pieces of timber and woody garden waste can be shredded for use in chipboard, or as a mulch for gardens or parks.

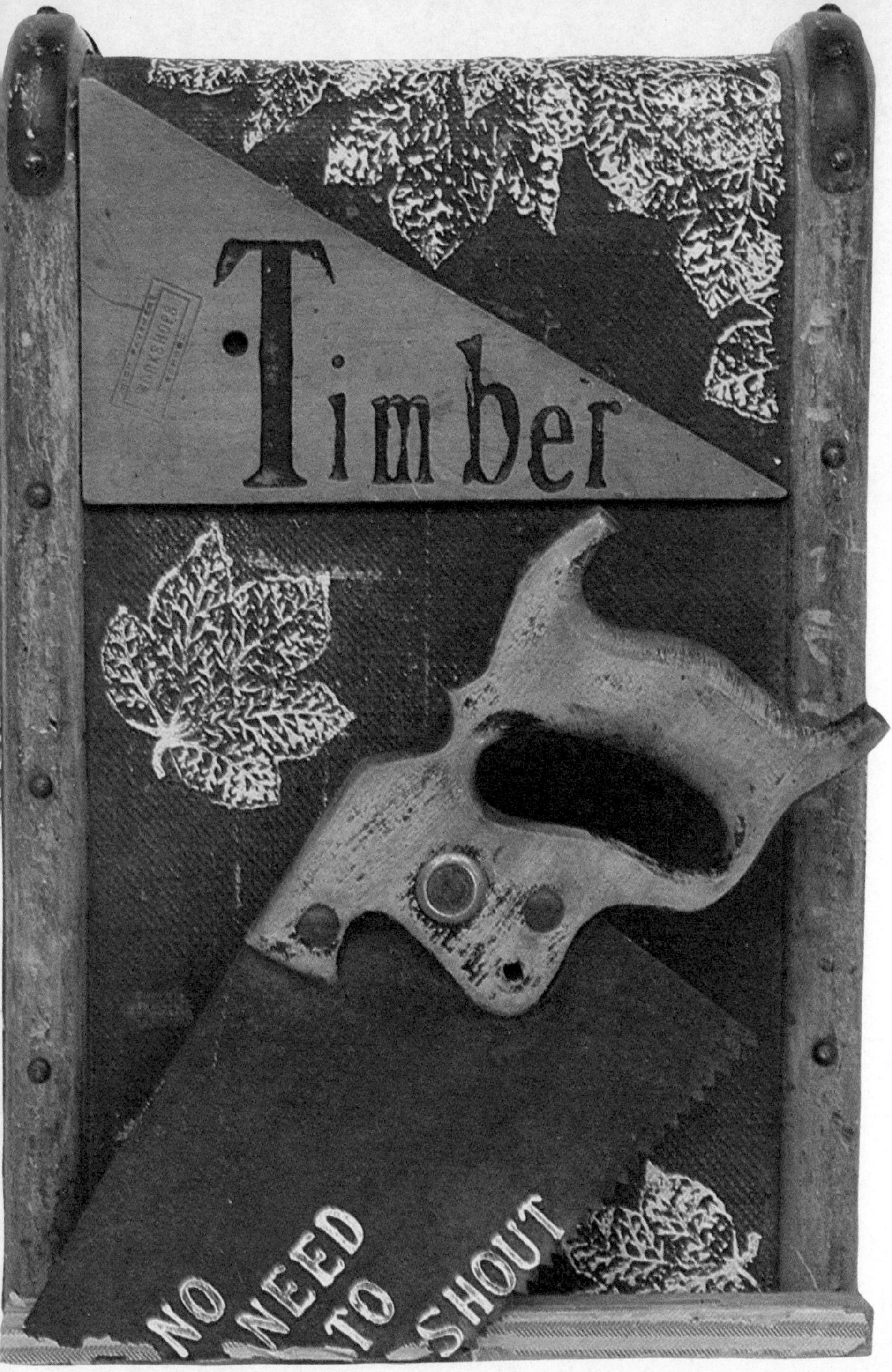
Timber
NO NEED TO SHOUT

16. TYRES

Vast numbers of used tyres are disposed of every year in the UK. Most of them are landfilled or just dumped.

Fires which break out in tyre dumps can burn for months or even years, and are almost impossible to control. Burning tyres give off poisonous fumes and leave an oily residue which can pollute water. This residue contains substances which can cause cancer.

Old tyres which are of good quality can be retreaded and reused. Tyres can also be used for boat protection, as weights for silage clamps, or to build crash barriers and children's playgrounds.

The rubber content of tyres can be removed and made into 'crumb', which can be mixed with asphalt to produce a good road surfacing material.

Some companies are setting up incineration plants to generate energy from burning tyres. Incineration is not recycling and is likely to create pollution problems of its own.

17. WHAT YOU CAN DO

This book has shown that much of the waste we produce in our homes need not be thrown away. Much of it can be avoided, reused, or recycled.

Everybody has a part to play in making sure that this is achieved.

The Government needs to set targets for reducing waste, and it must make sure that local authorities have the money they need to invest in recycling.

Manufacturing companies must reduce waste from their production processes, and make sure that facilities exist to enable their products to be recycled.

So what can you do?

This section contains some suggestions for ways in which you can play a part in putting an end to the throwaway society.

■ Reduce
Think about ways in which you could cut down the amount of waste you produce, at home and at work or school.

Avoid buying over-packaged goods. Take a shopping bag with you and refuse unnecessary carrier bags.

Cut down on the amount of paper you use at home and work. Write on both sides of paper. Circulate memos and documents instead of photocopying them.

■ Reuse
Try to avoid using disposable cups, plates, cutlery and towels. Ask the canteen at work, school or college to provide reusable alternatives.

Buy food, drinks and toiletries in returnable containers instead of in one-trip bottles, jars or cans. Ask your supermarket if it plans to 'bring back the bring back'.

Give unwanted clothes, books and household goods to jumble sales or charity shops, and buy things from them too.

Use scrap paper for writing notes and messages.

■ Recycle

Find out from your Council what facilities exist in your area for collecting organic waste, paper, glass, cans, plastics, textiles, furniture, oil and wood - and use them. If there aren't enough facilities ask the Council for more.

Start making your own compost - on a heap or in a worm bin - and use it in your garden instead of peat.

Encourage your school or workplace to set up a paper recycling scheme. There are many waste paper merchants operating throughout the country who collect paper from commercial premises.

■ Buy Recycled Products

Remember that collecting waste for recycling is only one step in a cycle. It is vital that the collected materials are reprocessed and used; so it is just as important to buy recycled products as it is to make sure your waste is collected for recycling.

Look out for recycled paper products made from low grade waste paper and bags made from recycled plastic.

FURTHER READING

Friends of the Earth publishes a large number of leaflets, briefings and reports on a wide range of environmental issues. The following are a selection; for a full list send a stamped addressed envelope to: Publications List, Friends of the Earth, 26-28 Underwood Street, London N1 7JQ. You can order the following books using the order form on page 50.

LBG165   **How to be a friend of the Earth** £3.45
An ideal guide for everyone who is worried about the planet and wants to know what they can do to help.

Informative, illustrated leaflets on:

LBG75	**Waste**
LBG124	**Recycling**
LBG206	**Reusable Packaging**
LBG225	**Water Pollution**
LBG43	**Air Pollution**
LBG42	**Acid Rain**
T63	**Countryside and Agriculture**
LBG62	**Peat**
LBG83	**Energy**
LBG166	**Energy Efficiency**
L91	**Nuclear Power**
L84	**Renewable Energy**
T107	**Rainforest Destruction and Third World Debt**
T109	**Tropical Rainforests** *50p each*

Recycling - Right Up Your Street
PR208   **wallchart** *only* £3.50
TP223   **teaching pack** £5.00
A full colour A1 wallchart illustrating a door-to-door recycling collection scheme, available with teacher's notes, suitable for Key Stage 3 and upper primary children.

REC159   **Recycling Officer's Handbook** £11.95
This popular handbook provides an overview of the political and legal contexts to recycling and a wealth of practical information on the options for collecting and sorting household and commercial waste.

L156   **Paper and the Environment** £1.00
This briefing explains the environmental problems caused
by the production and disposal of paper and outlines some of the
steps we can take to reduce these problems.

L205   **Bring Back the Bring Back** £1.50
This briefing examines the environmental benefits of reusable
packaging. It explores the reasons why so much disposable
packaging is used in the UK and proposes some solutions.

REC220   **Recycling City and Beyond** £11.95
This report looks at the practical initiatives
tested in Cardiff, Devon, Dundee and Sheffield as part
of Friends of the Earth's three-year Recycling City project,
and the lessons the project offers for national policy and local
action.

LBG129   **Paperchase - A Guide to Office Paper Collections** 50p
A guide to setting up an office waste paper collection
scheme.

TED170   **Lucy's World** £4.00
This magnificently illustrated book by Steve Weatherill looks
at the environment through the eyes of Lucy Goose and her
friends, bringing the issues alive for 5 to 7 year olds.

LBGP149   **Recycling Robot** £2.00
This poster provides information on recycling in a colourful
and fun way. It shows children how to make their own robot
collection unit and put recycling into practice.

L67   **Gardening Without Peat** £5.95
An easy reference book for gardeners on the alternatives to
peat.

L201   **21 Years of Friends of the Earth** £2.50
This booklet celebrates some of Friends of the Earth's key
campaign achievements since 1971.

Please use order form on page 50

FRIENDS OF THE EARTH
PUBLICATIONS ORDER FORM

HOW TO ORDER Please complete this order form and send with payment to:
Publications Despatch, Friends of the Earth, 26-28 Underwood Street, London N1 7JQ.

If you have any queries we will be happy to help you on 071 490 1555 ext.330.

NB Please write your order clearly giving both title and code number and enclose payment with your order.

Add on £2 for postage outside the UK (other than EEC)
Postage and packing is included in all prices
Please allow 28 days for delivery

Name Mr/Mrs/Ms/Miss

Address (for delivery)

Postcode   **Daytime telephone number**

Code	Title	Price	Quantity	Total

		Postage and Packing	Inclusive
		Donation	
		TOTAL	

Please make cheques payable to FRIENDS OF THE EARTH or complete your credit card details here: ☐ **Access**   ☐ **Visa**

				expires	

Cardholder's Signature

Cardholder's address (if different from above) _____

---

*If you do not want to cut out this coupon, please use a photocopy.

I WOULD LIKE TO SUPPORT THE WORK OF FRIENDS OF THE EARTH'S TRUST

Here is my donation of

☐ £100   ☐ £50

☐ £35   ☐ £15

Sum of your choice £ _____

I enclose total £ _____

payable to: *Friends of the Earth Trust Ltd* or debit my Access / Visa no:

Signature _____ Expiry date _____ / _____

Date _____ / _____ / _____

Name _____

Address _____

_____

_____ Postcode _____

Please send an SAE to the Membership Department if you would like information on any of the following:

☐ Legacies

☐ Covenants

☐ Membership of Friends of the Earth Limited

Phone to donate anytime 0582 485 805

Send to: **Friends of the Earth, 56-58 Alma Street, Luton, Beds LU1 2YZ.**

If you do not want to cut out this coupon, please use a photocopy.

PB 92092248